Weight loss and Dieting

By

Jose H. Lopez

TABLE OF CONTENT

Introduction

Welcome to "Weight Loss and Dietary: A Comprehensive Guide to Achieving Your Ideal Weight and Living a Healthy Lifestyle." In this book, we will embark on a transformative journey together, focusing on two crucial aspects: the importance of weight management and setting realistic goals with a positive mindset for successful weight loss.

Maintaining a healthy weight is not just about appearance; it profoundly impacts our overall health and well-being. Excess weight can increase the risk of various chronic conditions, such as heart disease, diabetes, and joint problems. By understanding the significance of weight management, we can take control of our health and work towards a better future.

Setting realistic goals is a pivotal step on the path to successful weight loss. It is essential to approach this journey with a positive mindset and embrace a sustainable approach rather than seeking quick fixes or drastic measures. By establishing attainable goals that align with our individual circumstances and lifestyle, we can lay the foundation for long-lasting, positive change.

Throughout this book, we will explore various facets of weight loss and dietary choices, providing you with the tools and knowledge necessary to make informed decisions about your health. We will start by understanding the science behind weight loss, debunking common myths and misconceptions that can hinder progress. Exploring the factors influencing weight loss, including genetics, lifestyle, and hormonal balance, will empower you to tailor your approach to suit your unique circumstances.

Assessing your current health and habits is a crucial step in understanding the areas that may be hindering your progress. We will guide you through a personal health assessment, enabling you to identify unhealthy eating patterns, lifestyle habits, and emotional triggers that contribute to weight gain. By gaining this self-awareness, you can implement effective strategies to overcome these obstacles.

Nutrition plays a fundamental role in weight management, and we will delve into the essentials of a healthy diet. Understanding macronutrients and their role in weight management, creating balanced and sustainable meal plans, and implementing portion control strategies and mindful eating practices will all be covered in detail.

Popular diets and weight loss programs often flood the market, leaving many feelings overwhelmed and uncertain about the best approach. In this book, we will provide an overview of various popular diets, such as keto, paleo, and intermittent fasting, and evaluate their effectiveness. Armed with this knowledge, you can make informed decisions about which approach suits your preferences and lifestyle best. Exercise is another critical component of weight loss, and we will explore its role and benefits. Designing a personalized exercise plan that incorporates cardio, strength training, and flexibility exercises will enable you to create a well-rounded fitness routine. We will also share tips for staying motivated and overcoming common exercise barriers that may arise.

Healthy lifestyle habits go hand in hand with weight loss. Managing stress and emotional eating, prioritizing quality sleep, and incorporating healthy habits into daily routines will be addressed to support your weight loss journey holistically.

Throughout this book, we will address common challenges that may arise, including weight loss plateaus, setbacks, cravings, and temptations. By building resilience and equipping yourself with effective strategies, you will be empowered to overcome these obstacles and maintain long-term weight loss success.

Finally, we will focus on sustainable weight maintenance and lifestyle transformation. Transitioning from a weight loss phase to a sustainable lifestyle is crucial to prevent yo-yo dieting and weight regain. We will discuss strategies to achieve weight maintenance, foster a positive body image, and cultivate a healthy relationship with food.

By embracing the principles and knowledge shared in this book, you will celebrate your progress and embark on a journey towards a healthier and more fulfilling life. Our aim is to provide you with final tips and advice for long-term success in weight management, ensuring that you continue prioritizing your health and well-being. Let's embark on this transformative journey together and unlock your potential for lasting change.

Chapter 1

Understanding Weight Loss

Section 1: Debunking Common Myths and Misconceptions about Weight Loss

When it comes to weight loss, there are numerous myths and misconceptions that can cloud our understanding and hinder our progress. In this chapter, we will debunk these myths and shed light on the truth behind successful weight loss.

1. Myth: "You can spot reduce fat in specific areas." Explanation: One common misconception is the belief that targeted exercises or specific diets can reduce fat in specific areas of the body. In reality, weight loss occurs as a result of overall fat reduction throughout the body. While specific exercises can strengthen and tone certain areas, they do not directly lead to spot reduction.

2. Myth: "Carbohydrates are the enemy." Explanation: Carbohydrates often get a bad reputation in the weight loss realm, but the truth is that not all carbs are created equal. It is important to distinguish between refined carbohydrates, such as sugary snacks and white bread, and complex carbohydrates, such as whole grains and fruits. The key is to focus on consuming complex carbohydrates in moderation as part of a balanced diet.

3. Myth: "Skipping meals accelerates weight loss." Explanation: Contrary to popular belief, skipping meals does not contribute to healthy weight loss. In fact, it can have negative

effects on your metabolism, energy levels, and overall well-being. Regular and balanced meals provide the necessary nutrients for your body's functions and help maintain stable blood sugar levels.

Section 2: The Science behind Calories, Metabolism, and Fat Burning

1. Understanding Calories: To comprehend weight loss, we must first understand the concept of calories. Calories are a measure of energy. When we consume more calories than our body needs, the excess is stored as fat, leading to weight gain. On the other hand, creating a calorie deficit through a combination of diet and exercise can prompt the body to burn stored fat for energy, resulting in weight loss.

2. Metabolism and Weight Loss: Metabolism refers to the processes in our body that convert food into energy. It plays a crucial role in weight loss because it determines how efficiently our bodies burn calories. Factors such as age, sex, muscle mass, and genetics can influence our metabolic rate. While we cannot control some of these factors, we can adopt lifestyle habits that support a healthy metabolism, such as regular exercise and balanced nutrition.

3. Fat Burning: Fat burning is a natural process that occurs when the body utilizes stored fat for energy. This process is triggered when there is an energy deficit, typically through a combination of reduced calorie intake and increased physical activity. Understanding the mechanisms behind fat burning can help us make informed choices regarding our diet and exercise routines to optimize weight loss.

Section 3: Factors Affecting Weight Loss, including Genetics, Lifestyle, and Hormonal Balance

1. Genetics and Weight Loss: Genetics can influence our predisposition to gain or lose weight. While we cannot change our genetic makeup, understanding our genetic tendencies can help us make realistic goals and develop strategies that work best for our individual circumstances.

2. Lifestyle Factors: Our lifestyle choices have a significant impact on our weight and overall health. Factors such as dietary habits, physical activity levels, sleep patterns, stress management, and smoking can all influence weight loss outcomes. By adopting

healthy lifestyle habits, we can create an environment conducive to successful weight management.

3. Hormonal Balance: Hormones play a crucial role in regulating our metabolism, appetite and fat storage. Hormonal imbalances, such as those related to thyroid function, insulin resistance, or cortisol levels, can affect weight loss efforts. Understanding the influence of hormones on weight management can guide us towards seeking appropriate medical advice and making necessary lifestyle adjustments.

Chapter 2

Assessing Your Current Health and Habits

Section 1: Conducting a Personal Health Assessment

Before embarking on a weight loss journey, it is crucial to assess your current health status. A personal health assessment provides valuable insights into your overall well-being, enabling you to identify areas that may need attention and improvement.

1. **Physical Health Assessment**: Conduct a comprehensive evaluation of your physical health. This may involve measuring vital signs such as blood pressure, heart rate, and body measurements (weight, height, waist circumference). Assessing your current health status will provide a baseline for tracking progress and identifying potential risk factors associated with excess weight.

2. **Medical History Review**: Take the time to review your medical history, including any pre-existing conditions or medications you are taking. Certain medical conditions can affect weight management and require special considerations. Discuss any concerns with your healthcare provider to ensure a safe and effective weight loss plan.

Section 2: Identifying Unhealthy Eating Patterns and Lifestyle Habits

1. Food Journaling: Keeping a food journal is a powerful tool for identifying unhealthy eating patterns. Track your daily food intake, including portion sizes and snacking habits. This practice promotes self-awareness and helps you identify patterns, such as emotional eating, mindless snacking, or reliance on processed foods.

2. Assessing Dietary Choices: Evaluate the nutritional quality of your diet. Are you consuming a variety of whole foods, including fruits, vegetables, lean proteins, and whole grains? Identify any imbalances or excessive consumption of certain food groups. This assessment will guide you in making healthier choices and creating a more balanced meal plan.

3. Lifestyle Habits Assessment: Assess your daily routines and habits to identify areas that may hinder your weight loss efforts. Consider factors such as physical activity levels,

sedentary behaviors, sleep patterns, and stress management techniques. Recognize any habits that may contribute to weight gain, such as a lack of exercise, excessive screen time, or inadequate sleep.

Section 3: Assessing Emotional Triggers and Their Impact on Weight Gain

1. Emotional Eating: Emotional eating is a common challenge that can derail weight loss efforts. Assess your relationship with food and identify emotional triggers that lead to overeating or unhealthy food choices. Reflect on situations, emotions, or stressors that prompt you to turn to food for comfort or distraction. Understanding these triggers will allow you to develop healthier coping mechanisms.

2. Stress and Weight Gain: Stress can significantly impact our weight. Assess the stress levels in your life and recognize the ways in which stress may contribute to emotional eating, disrupted sleep, or decreased physical activity. Identify effective stress management techniques that work for you, such as exercise, meditation, or engaging in hobbies.

3. Support Systems: Assess your support systems and the influence they have on your weight loss journey. Surround yourself with individuals who support your goals and provide encouragement. Consider seeking professional support from a registered loss support group to enhance your chances of success.

Chapter 3

Nutrition Essentials for Weight Loss

Section 1: Understanding Macronutrients and Their Role in Weight Management

Nutrition plays a vital role in weight management, and understanding macronutrients is key to creating a successful weight loss plan. In this chapter, we will delve into the importance of carbohydrates, proteins, and fats, and how they contribute to achieving and maintaining a healthy weight.

1. **Carbohydrates**: Carbohydrates provide energy for our bodies and come in two forms: simple and complex. Understanding the difference between the two can help you make informed choices. Complex carbohydrates, found in whole grains, fruits, and vegetables, provide essential nutrients and fiber, while simple carbohydrates, such as refined sugars, should be consumed in moderation. Balancing your carbohydrate intake is crucial for sustained energy levels and controlling blood sugar levels.

2. **Proteins**: Proteins are the building blocks of our bodies and play a significant role in weight management. They provide satiety, helping to reduce cravings and promote a feeling of fullness. Incorporating lean sources of protein, such as chicken, fish, tofu, legumes, and Greek yogurt, into your meals can support muscle maintenance and repair while aiding in weight loss.

3. **Fats**: Contrary to popular belief, fats are essential for a healthy diet. They provide energy, support hormone production, and help absorb fat-soluble vitamins. Opt for healthier fats, such as those found in avocados, nuts, seeds, and olive oil. Limit saturated and trans fats found in fried foods, processed snacks, and high-fat meats. Moderation and balance are key when including fats in your diet.

Section 2: Creating a Balanced and Sustainable Meal Plan

1. Establishing Caloric Needs: Determining your caloric needs is crucial for weight loss. Assess your activity level, age, gender, and weight goals to calculate a daily calorie

target. It is important to create a calorie deficit by consuming fewer calories than your body needs while still meeting your nutritional requirements.

2. Balancing Macronutrients: Develop a meal plan that includes a balance of carbohydrates, proteins, and fats. Focus on whole, nutrient-dense foods to provide the necessary vitamins, minerals, and fiber while promoting satiety. Incorporate a variety of fruits, vegetables, lean proteins, whole grains, and healthy fats into your meals.

3. Meal Prepping and Planning: Meal prepping and planning can save time and make healthier choices more accessible. Prepare meals and snacks in advance, ensuring they align with your caloric and nutritional needs. This practice helps prevent impulsive and unhealthy food choices when time is limited.

Section 3: Portion Control Strategies and Mindful Eating Practices

1. Portion Control: Portion control is essential for managing caloric intake. Use portion control techniques such as measuring cups, food scales, and visual cues to understand appropriate serving sizes. Be mindful of portion distortion, especially when dining out, and aim to consume balanced portions of each macronutrient.

2. Mindful Eating: Practicing mindful eating cultivates a healthy relationship with food. Slow down and savor each bite, paying attention to hunger and fullness cues. Mindful eating helps prevent overeating, allows for better digestion, and promotes enjoyment of the eating experience.

3. Emotional Eating Awareness: Develop an awareness of emotional eating triggers and strategies to cope with them. Find alternative ways to manage stress, boredom, or other emotions that may lead to mindless eating. Engage in activities such as journaling, practicing relaxation techniques, or engaging in hobbies to address emotional needs without turning to food.

Chapter 4

Popular Diets and Weight Loss Programs

Section 1: An Overview of Popular Diets and Their Effectiveness

In the world of weight loss, numerous diets and weight loss programs have gained popularity. In this chapter, we will explore some of the most popular diets, including the keto diet, paleo diet, and intermittent fasting, and assess their effectiveness in achieving weight loss goals.

1. The Keto Diet: The ketogenic diet is a low-carbohydrate, high-fat diet that aims to shift the body into a state of ketosis. It promotes the consumption of healthy fats while significantly reducing carbohydrate intake. We will discuss the science behind ketosis, its potential benefits for weight loss, and considerations for long-term adherence.

2. The Paleo Diet: The paleo diet focuses on consuming foods that mimic those of our ancestors from the Paleolithic era. It emphasizes whole, unprocessed foods and eliminates grains, legumes, and dairy. We will examine the principles of the paleo diet, its potential impact on weight loss, and its suitability for different individuals.

3. Intermittent Fasting: Intermittent fasting involves cycling between periods of fasting and eating. This approach can take various forms, such as the 16:8 method or alternate-day fasting. We will explore the potential benefits of intermittent fasting for weight loss, its impact on metabolism, and considerations for incorporating it into your lifestyle.

Section 2: Choosing the Right Diet Plan Based on Individual Preferences and Lifestyle

1. Individual Preferences: Choosing the right diet plan requires considering personal preferences, including food choices, eating patterns, and dietary restrictions. We will discuss how to align your preferences with the principles of different diets, allowing you to find a plan that suits your taste preferences and cultural background.

2. Lifestyle Factors: Consider your lifestyle when selecting a diet plan. Factors such as work schedule, social engagements, and cooking skills can influence your ability to adhere to certain diets. We will explore strategies for adapting popular diets to fit your lifestyle and provide tips for staying consistent and motivated.

Section 3: Evaluating the Pros and Cons of Commercial Weight Loss Programs

1. Commercial Weight Loss Programs: Commercial weight loss programs offer structured plans, support systems, and resources to aid in weight loss. We will examine popular programs such as Weight Watchers, Nutrisystem, and Jenny Craig, and assess their effectiveness, costs, and long-term sustainability.

2. Pros and Cons: Evaluate the pros and cons of commercial weight loss programs. Consider factors such as convenience, accountability, meal options, and potential drawbacks such as high costs or reliance on pre-packaged meals. This evaluation will help you make an informed decision about whether a commercial program aligns with your needs and preferences.

Chapter 5
Exercise for Weight Loss

Section 1: Exploring the Role of Physical Activity in Weight Loss

Physical activity plays a crucial role in achieving and maintaining weight loss. In this chapter, we will explore the benefits of exercise for weight management and overall well-being. Understanding the relationship between physical activity and weight loss will motivate you to incorporate exercise into your weight loss journey.

1. **Calorie Expenditure**: Exercise helps create a calorie deficit by burning calories. We will discuss how different types of physical activity impact calorie expenditure and contribute to weight loss. Understanding the energy expenditure of various exercises will help you make informed choices about your workout routines.

2. **Metabolism and Fat Burning**: Regular exercise can boost your metabolism and promote fat burning. We will explore how physical activity influences metabolic rate and fat oxidation. Learn about the benefits of both cardiovascular exercises and strength training for maximizing calorie burn and fat loss.

Section 2: Designing a Personalized Exercise Plan

1. **Cardiovascular Exercises**: Cardio exercises are effective for burning calories and improving cardiovascular health. We will discuss various cardio activities, such as running, cycling, swimming, and dancing, and their impact on weight loss. Learn how to incorporate cardio exercises into your routine and determine the appropriate intensity and duration for your fitness level.

2. **Strength Training**: Strength training is essential for building lean muscle mass, which can increase metabolism and support weight loss. We will explore different strength training exercises, including bodyweight exercises, weightlifting, and resistance training. Discover the benefits of strength training and how to incorporate it into your exercise plan.

3. **Flexibility and Mobility Exercises**: Flexibility and mobility exercises are often overlooked but are vital for overall fitness and injury prevention. We will discuss the importance of stretching and incorporating exercises such as yoga or Pilates into your routine. These exercises enhance flexibility, improve posture, and aid in recovery.

Section 3: Tips for Staying Motivated and Overcoming Exercise Barriers

1. Goal Setting: Set realistic and achievable exercise goals that align with your weight loss objectives. We will discuss the importance of setting specific, measurable, attainable, relevant, and time-bound (SMART) goals. Setting goals helps track progress and provides motivation throughout your weight loss journey.

2. Finding Enjoyable Activities: Choose activities that you enjoy to increase adherence and long-term sustainability. Discover different forms of exercise, such as group classes, outdoor activities, or team sports, and find what brings you joy and satisfaction. Exercise should be fun and something you look forward to.

3. **Overcoming Barriers**: Identify and overcome common barriers that may hinder your exercise routine. We will discuss strategies for dealing with time constraints, lack of motivation, or physical limitations. Learn how to incorporate physical activity into your daily life, even in small ways, to overcome obstacles and maintain consistency.

Chapter 6

Healthy Lifestyle Habits

Section 1: Strategies for Managing Stress and Emotional Eating

Introduction: Stress and emotional eating can sabotage your weight loss efforts. In this chapter, we will explore strategies to manage stress effectively and develop a healthier relationship with food. By implementing these strategies, you can prevent emotional eating and promote overall well-being.

1. **Recognizing Triggers**: Identify the sources of stress and emotional eating in your life. We will discuss common triggers and help you become more aware of the emotional connections to food. By recognizing these triggers, you can take proactive steps to manage them.

2. **Stress Management Techniques**: Learn effective stress management techniques to reduce emotional eating. We will explore various strategies, such as deep breathing exercises, meditation, journaling, and engaging in hobbies or activities that promote relaxation. Implementing these techniques can help you cope with stress without resorting to food.

Section 2: The Importance of Quality Sleep and Its Impact on Weight Management

1. **Understanding the Connection**: Quality sleep plays a crucial role in weight management. We will delve into the science behind sleep deprivation and its impact on hunger hormones, metabolism, and overall health. Recognize the importance of prioritizing sleep to support your weight loss journey.

2. **Establishing Healthy Sleep Habits**: Explore strategies to improve sleep quality and duration. We will discuss creating a sleep-friendly environment, developing a consistent sleep schedule, and implementing relaxation techniques before bed. These habits will contribute to better sleep and support your weight management goals.

Section 3: Incorporating Healthy Habits into Daily Routines

1. **Hydration**: Discover the importance of staying hydrated for weight loss and overall health. We will discuss the benefits of drinking enough water, strategies to increase water intake, and tips for making hydration a habit in your daily routine.

2. **Mindful Snacking**: Mindful snacking involves being aware of your eating habits and making conscious choices. We will explore techniques to practice mindful snacking, such as listening to your body's hunger and fullness cues, choosing nutritious snacks, and avoiding mindless eating.

3. **Meal Prepping**: Meal prepping is an effective way to support healthy eating habits and weight management. We will discuss the benefits of meal prepping, tips for planning and preparing meals in advance, and strategies for incorporating balanced and nutritious meals into your busy schedule.

Chapter 7

Overcoming Weight Loss Plateaus and Challenges

Section 1: Understanding Weight Loss Plateaus and How to Break Through Them

Weight loss plateaus can be frustrating, but they are a common part of the journey. In this chapter, we will explore why plateaus occur and provide strategies to overcome them, allowing you to continue making progress towards your weight loss goals.

1. **Understanding Plateaus**: Learn why weight loss plateaus happen and the factors that contribute to them. We will discuss physiological and psychological reasons for plateaus, such as metabolic adaptations, changes in body composition, and diminishing motivation. Understanding these factors will help you navigate through plateaus effectively.

2. **Breaking Through Plateaus**: Discover strategies to break through weight loss plateaus and kick-start your progress again. We will explore techniques such as adjusting calorie intake, changing your exercise routine, incorporating high-intensity interval training (HIIT), and exploring new forms of physical activity. These approaches can help you overcome plateaus and resume your weight loss journey.

Section 2: Dealing with Setbacks, Cravings, and Temptations

1. **Managing Setbacks**: Setbacks are a natural part of any weight loss journey. We will discuss strategies for bouncing back from setbacks, such as missed workouts, indulgent meals, or periods of low motivation. Learn how to adopt a growth mindset, reframe setbacks as learning opportunities, and stay focused on your long-term goals.

2. **Handling Cravings and Temptations**: Cravings and temptations can derail your progress if not managed effectively. We will explore techniques to handle food cravings, such as distraction strategies, mindful eating, and incorporating healthier alternatives. Additionally, we will discuss strategies for dealing with social situations and peer pressure that may trigger temptations.

Section 3: Building Resilience and Maintaining Long-Term Weight Loss Success

1. **Cultivating Resilience**: Resilience is essential for long-term weight loss success. We wil explore ways to build resilience, including setting realistic expectations, practicing self-compassion, and focusing on non-scale victories. Learn how to overcome obstacles, bounce back from setbacks, and stay committed to your goals.

2. **Sustainable Lifestyle Changes**: To maintain weight loss in the long run, it is crucial to make sustainable lifestyle changes. We will discuss the importance of creating healthy habits, finding balance, and avoiding restrictive diets or extreme measures. Emphasize the importance of adopting a holistic approach to health that includes nourishing your body, engaging in enjoyable physical activities, and prioritizing self-care.

Chapter 8

Sustainable Weight Maintenance and Lifestyle Transformation

Section 1: Transitioning from a Weight Loss Phase to a Sustainable Lifestyle

Transitioning from a weight loss phase to a sustainable lifestyle is crucial for maintaining long-term success. In this chapter, we will explore strategies to ensure a smooth transition and lay the foundation for a healthy, balanced lifestyle.

1. **Mindset Shift**: Shift your focus from solely weight loss to overall well-being. Embrace the mindset of sustainable lifestyle changes rather than temporary fixes. We will discuss the importance of setting new goals, celebrating non-scale victories, and finding intrinsic motivation for maintaining a healthy lifestyle.

2. **Behavior Reinforcement**: Reinforce the healthy habits you developed during your weight loss phase. We will explore techniques such as habit tracking, accountability, and positive reinforcement to solidify these behaviors as part of your daily routine. By making these habits second nature, you increase your chances of long-term success.

Section 2: Strategies for Weight Maintenance, Avoiding Yo-Yo Dieting, and Preventing Weight Regain

1. **Balanced Nutrition**: Maintain a balanced approach to nutrition by focusing on whole, nutrient-dense foods. We will discuss strategies to ensure you continue to meet your nutritional needs while enjoying a variety of foods. Learn how to incorporate occasional indulgences without derailing your progress.

2. **Portion Control and Mindful Eating**: Continue practicing portion control and mindful eating as essential tools for weight maintenance. We will explore techniques such as listening to hunger and fullness cues, slowing down during meals, and savoring the eating experience. These practices promote a healthy relationship with food and prevent overeating.

3. **Regular Physical Activity**: Sustain an active lifestyle by finding activities you enjoy and incorporating them into your routine. We will discuss the importance of regular physical activity for weight maintenance and overall well-being. Explore different forms of exercise, set new fitness goals, and prioritize movement as an integral part of your daily life.

Section 3: Embracing a Positive Body Image and Fostering a Healthy Relationship with Food

1. **Positive Body Image**: Cultivate a positive body image and practice self-acceptance throughout your journey. We will explore techniques to foster a healthy body image, such as reframing negative self-talk, surrounding yourself with positive influences, and practicing self-care. Embrace your unique body and appreciate it for its strength and functionality.

2. **Intuitive Eating**: Develop a healthy relationship with food by practicing intuitive eating. We will discuss the principles of intuitive eating, such as honoring hunger and fullness, rejecting diet mentality, and finding satisfaction in food choices. By trusting your body's cues, you can create a sustainable and nourishing approach to eating.

Conclusion

Congratulations on completing the journey of exploring weight loss and dietary strategies for a healthier life! Throughout this eBook, we have covered various aspects of weight management, debunked common myths, and provided practical advice to support your goals. As you reach the end of this book, take a moment to reflect on your progress and the positive changes you've made.

1. Celebrating Progress and Embracing a Healthier and More Fulfilling Life: Remember to celebrate each milestone along your weight loss and dietary journey. Whether it's shedding pounds, adopting healthier habits, or experiencing improved well-being, every step forward is an achievement. Embrace the positive changes you've made, and let them fuel your motivation to continue on this path towards a healthier and more fulfilling life.

2. Final Tips and Advice for Long-Term Success in Weight Management: As you move forward, here are some important tips to help you maintain your achievements and ensure long-term success in weight management:

- Stay consistent: Consistency is key when it comes to maintaining a healthy lifestyle. Continue to prioritize nutritious eating, regular physical activity, and self-care practices.
- Set realistic goals: Set realistic and sustainable goals that align with your individual needs and preferences. Remember that health is not just about the number on the scale but also about overall well-being.
- Seek support: Surround yourself with a supportive network of friends, family, or even online communities that can provide encouragement and accountability. Consider seeking professional guidance from a registered dietitian or a health coach if needed.
- Stay adaptable: Life is full of changes, and your weight management journey may require adjustments along the way. Embrace flexibility and be open to modifying your approach as needed to suit your evolving needs and circumstances.

3. Encouragement to Continue Prioritizing Health and Well-being: Above all, remember that prioritizing your health and well-being is an ongoing journey. It's not about achieving a specific weight or reaching a destination—it's about living a vibrant and

balanced life. Continue to make conscious choices that support your overall well-being, both physically and mentally. Embrace self-love, practice self-care, and be kind to yourself throughout the process.

As you close this eBook, let it serve as a reminder of the knowledge and insights you've gained. Apply these principles in your daily life and keep the flame of your motivation burning bright. Your health and well-being are worth the ongoing effort and commitment.

Here's to your continued success in weight management, embracing a healthier and more fulfilling life, and prioritizing your health and well-being. You have the power to transform your life, one step at a time.

www.ingramcontent.com/pod-product-compliance
Lightning Source LLC
Chambersburg PA
CBHW070916220526
45466CB00005B/2237

* 9 7 9 8 3 9 6 2 9 0 1 6 7 *